Education
93

地球的皮肤

The Earth's Skin

Gunter Pauli

［比］冈特·鲍利　著

［哥伦］凯瑟琳娜·巴赫　绘

田　烁　王菁菁　译

上海远东出版社

丛书编委会

主　任：田成川

副主任：何家振　闫世东　林　玉

委　员：李原原　翟致信　靳增江　史国鹏　梁雅丽
　　　　任泽林　陈　卫　薛　梅　王　岢　郑循如
　　　　彭　勇　王梦雨

特别感谢以下热心人士对童书工作的支持：

匡志强　宋小华　解　东　厉　云　李　婧　庞英元
李　阳　刘　丹　冯家宝　熊彩虹　罗淑怡　旷　婉
杨　荣　刘学振　何圣霖　廖清州　谭燕宁　王　征
李　杰　韦小宏　欧　亮　陈强林　陈　果　寿颖慧
罗　佳　傅　俊　白永喆　戴　虹

目录

Contents

一只几维鸟正在新西兰海岸边的灌木丛中觅食，他发出了求偶的叫声，试图吸引雌性同伴。这时，一只棕色小鸟飞了下来，他长着长长的喙，翅膀上有黑白相间的斑点。

"你不是斑尾塍鹬吗？"几维鸟问道，"你是刚从北半球的阿拉斯加一路飞到南半球的我们这儿来的吗？"

A kiwi is foraging in the bush along the shores of New Zealand. He gives his mating call, trying to attract a female, when a brown bird with a long beak and black and white spotted wings alights next to him.

"Are you not a godwit?" asks the kiwi. "And have you just come all the way from Alaska in the Northern Hemisphere to join us here in the Southern Hemisphere?"

你不是斑尾塍鹬吗？

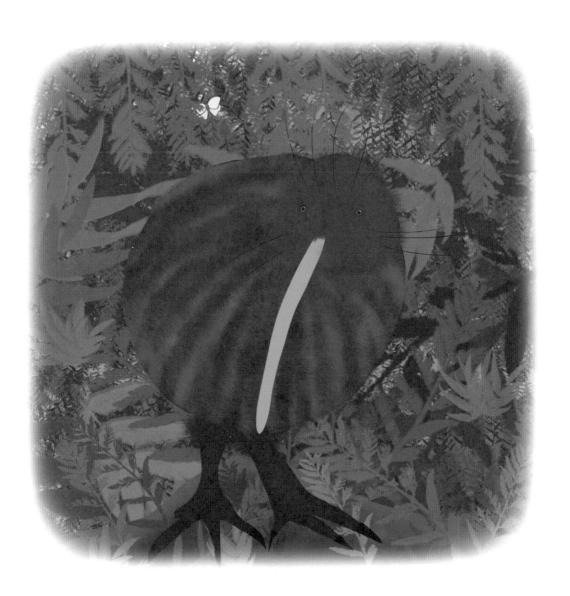

Are you not a godwit?

著名的没有翅膀的鸟

The famous wingless bird

"是的，我的确是从那里来的。你不是几维鸟吗？著名的没有翅膀的鸟？"斑尾塍鹬回应着。

"没错，我就是鸟儿家族中不会飞的一员。但是，尽管没有翅膀，我们也是非常快乐的。"

"Yes, I am. And that is exactly where I have come from. And are you not a kiwi, the famous wingless bird?" responds the godwit.

"Indeed. I am one of those exceptional birds that cannot fly. But even without wings, we are a happy lot."

"你看上去的确很快乐，而这是生命中最重要的事情。"

"我听过好多关于你的事情。没有任何一种鸟能一站不停地飞得比你更远，就算人类坐着他们制造的机器也不行。"

"嗯，是的，好像就连飞机都不能和我的飞行能力相比，我能一口气飞行18 000千米呢。"

"You do look happy. And that is the most important thing in life."

"I've heard so much about you. And how no one, not even people with their machines, can fly non-stop longer than you."

"Well, yes, it seems that even planes cannot yet match my ability to do a 18 000-kilometre long trip."

没有任何一种鸟能一站不停地飞得比你更远！

No one can fly non-stop longer than you!

你的意思是你们自己吃自己？

Does it mean you eat yourself?

"飞行中你们从来不停下来小便或吃点东西吗？"几维鸟问。

"我们在飞行中小便，在出发之前，我们身体重量的一半是脂肪，就是用来充饥的。"

"我听不懂你刚才说的话。你的意思是你们自己吃自己？"

"才不是呢！我的身体会将那些脂肪转化为能量，这样我就有力气远途飞行了。整个动物界都知道，我的飞行距离是最远的。"

"Do you never have to stop along the way to wee or eat?" the kiwi asks.

"We wee as we fly, and before leaving, more than half of our body weight is fat for food."

"I don't understand what you just said. Does it mean you eat yourself?"

"No! My body just turns that fat into energy so I have all the power I need to make the longest journey known in the animal world."

"你们斑尾塍鹬会吃枸杞、奇亚籽、螺旋藻之类的超级食品？"

"噢，不，我们就地取食。我们吃昆虫、蜗牛、贝壳和水草。"

"那一定是非常健康的饮食。"

"是的，但是我可以告诉你，从空中看，地球可没有这么健康。在陆地上空飞行，我可以看到地球的皮肤是如何渐渐干枯的。"

"Do you godwits eat superfoods like goji berries, chia seeds, and spirulina?"

"Oh no, we eat whatever is locally available. We eat insects, snails, shells, and water plants."

"That must be a very healthy diet."

"Yes, but I can tell you that seen from the sky, the Earth does not look so healthy. Flying over the land I can see how the earth's skin is shrivelling."

我们吃昆虫、蜗牛、贝壳和水草

We eat insects, snails, shells, and water plants

……每一处森林都遭到了砍伐……

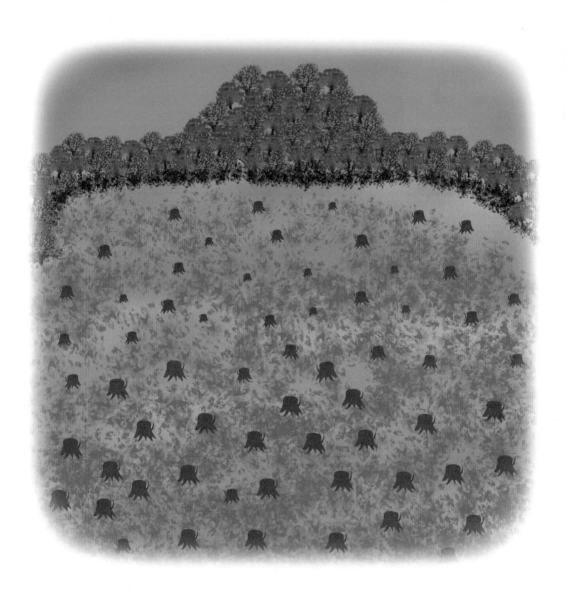

… forest had been cut down everywhere …

"你这话是什么意思？地球没有皮肤，只有人类和其他动物有皮肤。"几维鸟说。

"地球的皮肤是指覆盖地球表面的薄薄的土壤层，它让地球上的生命得以生存。我在飞行时看到，阿拉斯加的积雪正在融化，太平洋的岛屿正在萎缩。还有，当我飞越陆地时，我看到每一处森林都遭到了砍伐。"

"陆地正在下沉吗？"几维鸟问道。

"What do you mean? The planet does not have a skin; only people and animals have skin," says the kiwi.

"The skin is the thin layer of earth that covers the planet and allows for life to exist on Earth. When I fly, I see snow melting in Alaska, islands shrinking in the Pacific, and when I pass over land, I see that forests had been cut down everywhere."

"Is the earth sinking?" asks the kiwi.

"没有。水变得越来越热，并因此不断膨胀，水平面开始上升。但是，最糟糕的是，没有树木或适宜的植被，土壤变得越来越干旱。"

"人类焚烧土地了？"

"人类焚烧所有他们触手可及的东西：树木、垃圾，甚至还有食物、电脑及其他电子设备。"

"No, water is getting warmer and therefore expanding and then water levels rise. But the worst thing is that land without trees or without proper plant cover becomes dry."

"Do people burn the land?"

"People burn everything they can get their hands on: trees, waste, even food, and also computers and other electronic devices."

人类焚烧土地了?

Do people burn the land?

人类有一些不错的想法

People have some great ideas

"还有办法吗？你认为我们还有希望吗？"

"嗯，人类有一些不错的想法：比如从水流、太阳和风中获取能量。吃有机食物也是迈向正确方向的一步，但是……"

"这还远远不够，对吗？"几维鸟问道。

"Is there a way out? Do you think there any hope for us?"

"Well, people have some great ideas: generating power from the flow of water, sun, and wind; and eating organic food is also a step in the right direction, but …"

"That's not good enough, is it?" asks Kiwi.

"是的，"斑尾塍鹬答道，"我们亟需找到让当地原有植被重新覆盖土壤的方法。我们还要把流失的地表土壤复归原位，它们可是地球的皮肤啊！这样土地就能再次焕发生机了。"

"那样，我就能回到我的家乡了，在森林里，晚上出来觅食。"

……这仅仅是开始！……

"No," replies the godwit. "We urgently need to find ways to once again cover the land with the native plants that used to grow there. And we need to replace lost topsoil, which is the skin of the Earth, so that the land can power life again."
"And then I can return to where I belong, in the forest, foraging at night."
... AND IT HAS ONLY JUST BEGUN!...

······这仅仅是开始！······

... AND IT HAS ONLY JUST BEGUN! ..

你知道吗？

The distance of the longest non-stop commercial flight, using fuel, is approximately 13 800 km. The distance of the longest non-commercial flight, using fuel, is 18 000 km. The longest flight using only solar power covered a distance of 8 200 km.

在不经停的情况下，使用燃油的商业航班，最远飞行距离大约是13 800千米，使用燃油的非商业航班是18 000千米，纯太阳能动力航班是8 200千米。

= 13,800 千米 (✈)
= 18,000 千米 (✈)
= 8,200 千米 (☀)

The bar-tailed godwit flies non-stop from the Arctic to the coasts of Australia and New Zealand. This bird weighs 400 g when it starts its journey and on arrival weighs only 200 g. It undertakes the longest non-stop journey, lasting eight to nine days, without landing to feed.

斑尾塍鹬能直接从北极地区飞到澳大利亚和新西兰沿岸，不需停下来休息。这种鸟类在旅途开始时的体重为400克，到达目的地后仅为200克。这是最长的不间断旅途，通常需要飞上8到9天，中途不着陆进食。

8—9天

The bar-tailed godwit feeds on insects, crustaceans, and aquatic plants. The bird accumulates fat, which is converted into energy during the flight.

斑尾塍鹬以昆虫、甲壳动物和水生植物为食，它们囤积脂肪，在飞行途中将其转化为能量。

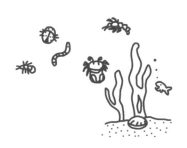

The Maori, the native people of New Zealand, call the migratory godwit the "mystery bird" because it does not nest in New Zealand and because young birds take four years before starting to breed.

新西兰的原住民族毛利人，将迁徙的斑尾塍鹬称作"神秘鸟"，因为它们不在新西兰筑巢，且幼鸟需要4年的时间才开始繁育后代。

When the godwit arrives at its destination it will first eat berries to replenish its energy.

斑尾塍鹬到达目的地后，先要吃一些浆果来补充能量。

能量

The kiwi is nocturnal, lives in the forest, and is about the size of a hen. It is the only bird that has nostrils, which it uses for sniffing out food. It has the largest egg-to-body weight ratio (15% compared to 2% for the ostrich), laying an egg six times bigger than a chicken egg.

几维鸟是夜行动物，生活在森林中，大小和母鸡差不多。它是唯一长有鼻孔并用鼻孔来闻味觅食的鸟类。它的蛋与体重比最大（为15%，而鸵鸟为2%），而且比鸡蛋要大6倍。

The kiwi has a long life span of 25 to 50 years. There are only 70 000 left in the wild and the population reduces by 2% every year.

几维鸟的寿命很长，有25—50年。野生几维鸟的数量只有7万只，且每年以2%的幅度减少。

The chemical composition of the atmosphere is the result of chemical reactions on the surface of the Earth. So, if we change the composition of the multilayers of the Earth, including the vegetation, then we change the atmosphere.

大气的化学成分是地球表面化学反应的结果。因此，如果我们改变了地球包括植被在内的多层结构，那么我们将改变大气的成分。

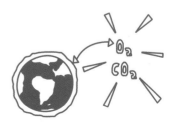

你相信吗？一只仅重400克的小鸟能够执行如此长距离的不间断飞行任务，距离之长远超人造飞行器。

Can you believe that a small bird weighing only 400 g can undertake such long flights and travel farther on a non-stop journey than any man-made aircraft?

你想不想从空中看看我们的世界？它看起来会很不一样吗？你能因此看到更多吗？

Would you like to see the world from the air? Would it look very different and would you be able to see more?

如果我们仅靠身体脂肪，不需要任何燃料或太阳能就能飞行，将会发生什么呢？

What if we were able to fly by only using our body fat and not needing any fuel or solar energy?

你能为改善地球的皮肤（地表土壤）做些什么？

What can you do to improve the Earth's skin (topsoil)?

Take a close look at your skin. The quality of your skin says a lot about your health. Skin problems can be the first sign of serious underlying health problems such as diabetes and lung cancer. So learn to monitor your health by looking at your skin. A butterfly rash (red spots on the face) may be the first sign of lupus. Velvety plaque on the skin suggests diabetes.

The key is to be aware of such conditions and to keep a close check on our skin. Like the bird observing the health of the earth by looking at the topsoil from the sky, observe what your skin is telling you.

　　仔细看看你自己的皮肤，肤质传递了很多关于健康状况的信息。皮肤问题是很多严重的潜在健康问题（比如糖尿病和肺癌）的最初症状。所以，要学会通过观察皮肤来掌握自己的健康状况：蝴蝶状皮疹（脸上的红色斑点）可能是红斑狼疮的预兆；皮肤上软软的斑块则提示可能患上了糖尿病。关键是要了解这些情况，并密切关注我们的皮肤。就像鸟儿通过在天上看地表土壤来观察地球健康状况一样，你要观察皮肤向你展示的身体状况。

学科知识
Academic Knowledge

生物学	几维鸟不会飞,因森林被滥砍滥伐改变了自然习性而濒临灭绝;几维鸟的夜间行动是因食肉动物的攻击改变了自然习性后的适应表现;几维鸟终生配对;斑尾塍鹬主要以脂肪形式储存能量;森林空巢或不育的新现象:因过度开发、无节制的狩猎和单一栽培,一片长满树木的森林中几乎没有任何动物。
化 学	人类的代谢作用不能将脂肪转化成必需的氨基酸或蛋白质;鸟类将脂肪用作燃料,因为每单位质量的脂肪中含有更多的能量;大气的产生和保持是通过地球土壤和植被的多层复杂反应来实现的。
物 理	脂肪更轻,因此比含有相同能量的其他食物更容易携带;森林有冷却作用:云层越过森林地区时更容易分解出水分而下雨。
工程学	工程师将焚烧技术作为一种常用的减少体积、回收热量和辅助分离的方法。
经济学	城市在森林中获取原材料,用于建筑、船只和陶瓷,这都是城市的发展陷阱,这种大规模的自然资源开发活动,加之采矿业和冶金业的发展,通常会引发巨大的经济危机;在13和14世纪,欧洲的森林砍伐活动达到了高峰,导致木材的短缺以及人口营养水平不断下滑;当时的欧洲通过从美洲地区进口土豆和玉米,才得以从食物短缺的灾难中存活下来。
伦理学	人们往往将自己认为没用的东西一烧了之,而不去试着发现这些东西更多的潜在价值。
历 史	古希腊因对木材的过度需求而遭受了森林滥砍滥伐的灾难;伊斯特岛因森林损毁而水土流失,这导致了文明的衰落;15世纪,因航道淤积,人口数量和食物需求的不断增加,以及无节制的森林砍伐活动,布鲁日港不得不迁往安特卫普(比利时)。
地 理	几维鸟是新西兰的国家象征;冰川融化导致的海平面上升对印度洋和太平洋的许多小岛国产生了影响。
数 学	如果斑尾塍鹬飞行18 000千米需要消耗200克脂肪,1千克脂肪完全氧化释放9 000千卡的热量,那么它就能够仅靠1 800千卡的热量绕地球飞行半圈。
生活方式	超级食物的兴起;不管什么垃圾,人们总是迫不及待地将其焚烧,通常不回收热量,而仅关注垃圾减量。
社会学	"几维鸟"一词源于毛利语,是鸟叫的象声词;几维果(奇异果、猕猴桃)原产于中国,最初叫作中国醋栗。
心理学	自我意识;没有达到或满足团队期望的时候也能感受到快乐的能力;如果你心情愉快,就更容易从疾病中康复。
系统论	滥砍滥伐导致生物多样性的丧失,加速水土流失,引发气候变化,改变大气的化学成分;生命赖以生存的薄薄的土壤表层是一个复杂的系统,它由冠层、林下叶层、灌木层、草本层、地表覆盖层等组成。

情感智慧
Emotional Intelligence

几维鸟

几维鸟善于社交，愿意和陌生人对话。他有自我意识，知道自己是规律中的一个例外，但很快指出自己是快乐的。几维鸟向斑尾塍鹬表达了钦佩之意，渴望了解他如此远距离飞行的能量来源，却发现那只不过是普通的简单食物。因砍伐森林，几维鸟被迫离开了原来的栖息地，听到关于"地球皮肤"的表述，他感到很惊讶，也很困惑。这就促使他提出了一系列问题，并畅想回到森林家园的美好未来。

斑尾塍鹬

斑尾塍鹬发起了对话，他承认了自己的身份，快速回答了问题，并想知道和他对话的这只鸟是否就是著名的几维鸟。斑尾塍鹬也有自我意识，充分认识到自己的良好表现，但并没有表现出自私和傲慢。相反，他非常务实，大方地分享自己的主要饮食特点、代谢和消化系统，以及排泄方式。他们的对话非常清晰，观点十分明确。然而，斑尾塍鹬将注意力转到了地球正面临的一些问题。当几维鸟质疑"皮肤"这个词时，斑尾塍鹬没有直接回答，而是阐述了一些他飞越半个地球所领略到的地球正面临的状况。最后他提出了一个清晰的解决方案：再造森林，让地球恢复原来的样子。

艺术
The Arts

从太空中看，地球是非常美丽的，我们一起来绘制一个地球吧！在一个任意大小的球上面绘图。因为球体没有能让你握住的棱角，所以要找一些速干的颜料。在球上画出陆地和海洋，在陆地上尽可能多地覆上森林。然后，给它拍一张照片，发到互联网上，让其他人也能欣赏你的作品。

思维拓展
Systems: Making the Connections

　　人类当前越来越重视对濒危物种的保护，这没错，但是我们似乎失去了更广阔的视野。地球上的生命被圈定在一个薄薄的多层结构之内，包括土壤、水、植被以及一层由动物、真菌和细菌组成的致密层，核心的化学反应正是发生在这些仅几米厚的系统中。如果我们改变了地球的样貌，那么我们也改变了大气的组成。因为森林覆盖率已经减少了一半，我们不再使用那些丰富、高效、可再生的空间环境了，而是改为种植单一作物，比如水果、蔬菜、树木。这就只有植被生长，而没有真菌、藻类和动物的存在。我们已经打破了生命之间的微妙关系，这种共生的生态系统已经发展了很久。也许最重要的改变莫过于地球正在急速丧失自己的表层土壤。它富含碳元素，是许多养料的来源，却被过度开发，以至于碳储量几乎消失殆尽。而且，地球的天然自我修复能力也被破坏了。尽管全世界范围内的传统文明曾经发明过很多修复植被、恢复土壤的方法，但这些技术几乎都被丢弃了。中国很早就有修复干旱土地的技术——将桑树种植和养蚕相结合，这样，蚕的粪便就可以让土壤重新恢复肥力。现在，人们已经越来越意识到，需要将大自然的五界整合并再生，并开始将废弃的土地复垦为耕地。

动手能力
Capacity to Implement

　　仔细观察一下你的学校和它的周围环境。你如何恢复你们学校及其周围的生态系统，使表层土壤再生并生产粮食？和你的老师甚至校长聊聊这件事。你可以问以下问题：所有树的树冠都可以产生树荫吗？哪种植物可以形成林下叶层？校园里有多少灌木丛？有没有一种草本植物可以用来烹饪和制作草药？哪些植物可以固定氮？土壤中植物根系周围的区域健康状况如何？那里应该有很多的真菌、昆虫、线虫和蠕虫。最后一个问题，你有没有看到一些横向或纵向生长的攀援植物？你可以设计一个方案，让每一层的生命物质都存在，让大气中化学成分重新恢复正常。

故事灵感来自
This Fable Is Inspired by

比尔·莫里森
Bill Mollison

比尔·莫里森1928年出生在澳大利亚的塔斯马尼亚岛，被称为"永续农业之父"。他与戴维·霍姆格伦一起提出了"永续农业"这一概念，还共同出版了《永续农业》一书。永续农业通过模拟自然系统来界定农业和社会发展，基于关爱地球、关爱人类、多者退还的原则之上。其哲学思想是与自然协同工作，而不是对立；关注所有动植物的功能及其相互关联，而不仅是一种功能。比尔相信，他可以教会任何人成为园丁，一旦他们知道了如何耕作，他就可以将这些人变为哲学家。但是，他不能先把一个人教成哲学家，然后再将他变成一名园丁。

图书在版编目（CIP）数据

冈特生态童书.第三辑修订版：全36册：汉英对照 /
（比）冈特·鲍利著；（哥伦）凯瑟琳娜·巴赫绘；
何家振等译.—上海：上海远东出版社，2022
书名原文：Gunter's Fables
ISBN 978-7-5476-1850-9

Ⅰ.①冈… Ⅱ.①冈… ②凯… ③何… Ⅲ.①生态环
境–环境保护–儿童读物—汉、英 Ⅳ.①X171.1-49

中国版本图书馆CIP数据核字（2022）第163904号
著作权合同登记号图字09-2022-0637号

策　　划　张　蓉
责任编辑　祁东城
封面设计　魏　来李　廉

冈特生态童书
地球的皮肤
[比]冈特·鲍利　著
[哥伦]凯瑟琳娜·巴赫　绘
田　烁　王菁菁　译

记得要和身边的小朋友分享环保知识哦！
八喜冰淇淋祝你成为环保小使者！